P9-DBS-478

WHEELS
AT WORK AND PLAY

ALL ABOUT DIGGERS

Wheels at Work and Play
All about Diggers
All about Motorcycles
All about Race Cars
All about Special Engines
All about Tractors
All about Trucks

Library of Congress Cataloging-in-Publication Data

Stickland, Paul.
All about diggers / Paul Stickland.
p. cm. — (Wheels at work and play)
Summary: Depicts different kinds of earth moving machines and their various uses.
ISBN 0-8368-0422-8
1. Earthmoving machinery—Juvenile literature. [1. Earthmoving machinery. 2. Machinery.] I. Title. II. Series.
TA725.S79 1990
629.225—dc20 90-9820

This North American edition first published in 1990 by
Gareth Stevens Children's Books
1555 North RiverCenter Drive, Suite 201
Milwaukee, Wisconsin 53212, USA

Series editor: Tom Barnett
Designer: Laurie Shock

Printed in the United States of America

1 2 3 4 5 6 7 8 9 96 95 94 93 92 91 90

WHEELS
AT WORK AND PLAY

ALL ABOUT DIGGERS

Paul Stickland

Gareth Stevens Children's Books
MILWAUKEE

The huge bulldozer flattens the ground.

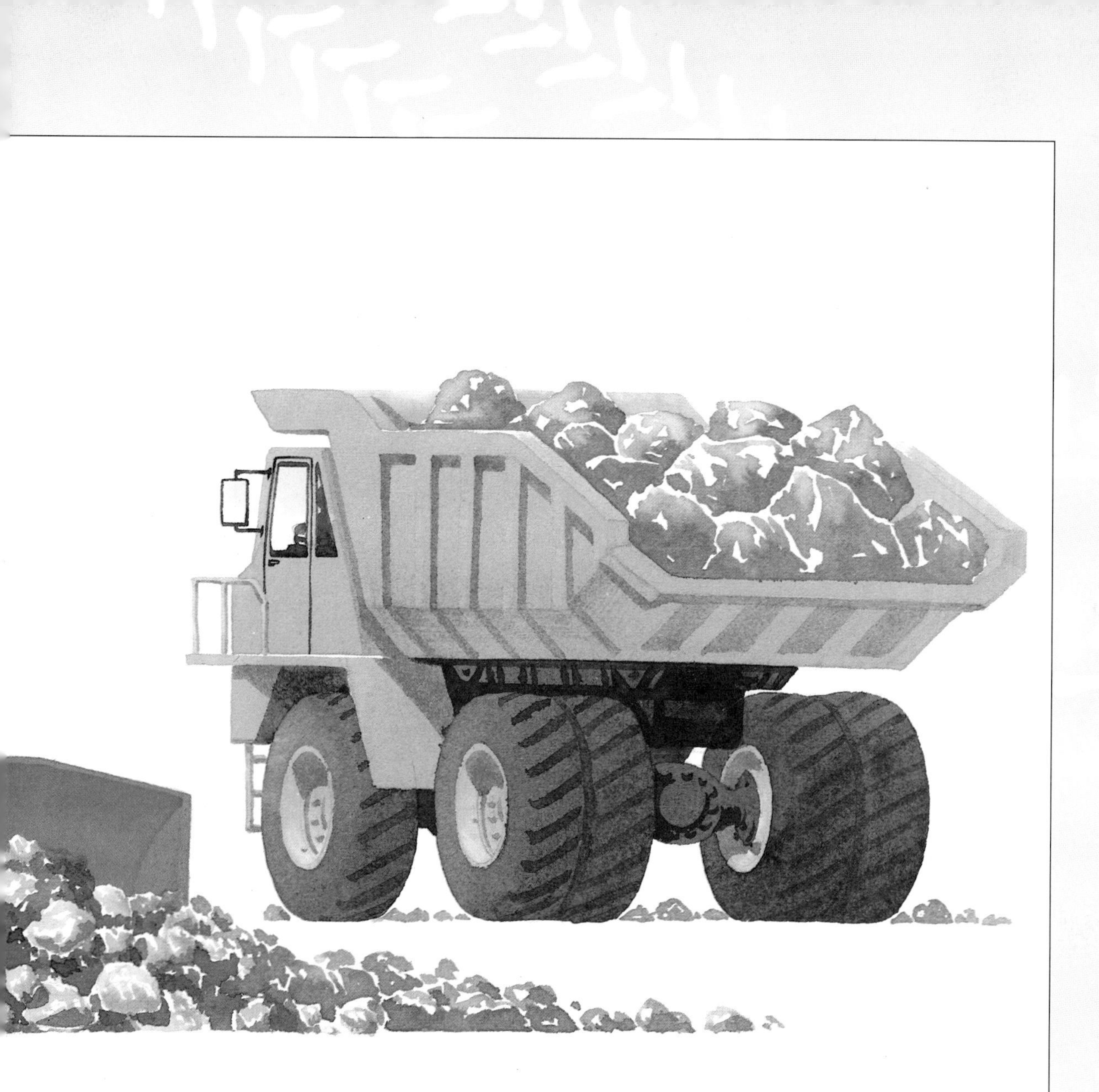

The giant dump truck
carries rocks.

This machine faces forward and backward.

It is a digger and bulldozer.

This is an excavator.
It digs holes.

It crawls along on
Caterpillar treads.

The yellow loader has
scooped up rocks.

It fills the red dump truck.

This bulldozer is heavy.

It must be moved on a special transporter.

Small and quick machines are sometimes useful.

This one fills a dump truck
with earth.

Glossary

bulldozer
A truck that has a large scoop in the front which can move large rocks.

Caterpillar treads
Special wheels on a truck that help it drive over rough ground.

dump truck
A truck that can tip its back and dump its contents.

excavator
A digging truck with a large shovel.

transporter
A long, flat truck used to carry other trucks.

Index

AUG 16 1991 MAY 25 2000
J629.225
S C 3
Stickland
All about diggers
DISCARD
SANFORD, N.C.